BUILDING BLOCKS OF CHEMISTRY

CHEMISTRY AND MATTER

Written by Cassie Meyer

Illustrated by Maxine Lee-Mackie

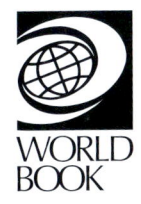

WORLD BOOK

a Scott Fetzer company
Chicago

World Book, Inc.
180 North LaSalle Street
Suite 900
Chicago, Illinois 60601
USA

For information about other World Book publications, visit our website at **www.worldbook.com** or call **1-800-WORLDBK (967-5325)**.
For information about sales to schools and libraries, call 1-800-975-3250 (United States), or 1-800-837-5365 (Canada).

© 2023 World Book, Inc. All rights reserved. This volume may not be reproduced in whole or in part in any form without prior written permission from the publisher.

WORLD BOOK and the GLOBE DEVICE are registered trademarks or trademarks of World Book, Inc.

Library of Congress Cataloging-in-Publication Data for this volume has been applied for.

Building Blocks of Chemistry
ISBN: 978-0-7166-4371-5 (set, hc.)

Chemistry and Matter
ISBN: 978-0-7166-4376-0 (hc.)

Also available as:

Printed in India by Thomson Press (India) Limited, Uttar Pradesh, India
1st printing June 2022

WORLD BOOK STAFF
Executive Committee
President: Geoff Broderick
Vice President, Editorial: Tom Evans
Vice President, Finance: Donald D. Keller
Vice President, Marketing: Jean Lin
Vice President, International: Eddy Kisman
Vice President, Technology: Jason Dole
Director, Human Resources: Bev Ecker

Editorial
Manager, New Content: Jeff De La Rosa
Associate Manager, New Product: Nicholas Kilzer
Sr. Content Creator: William D. Adams
Proofreader: Nathalie Strassheim

Graphics and Design
Sr. Visual Communications Designer: Melanie Bender
Sr. Web Designer/Digital Media Developer: Matt Carrington

Acknowledgments:
Writer: Cassie Meyer
Illustrator: Maxine Lee-Mackie/ The Bright Agency
Series Advisor: Marjorie Frank
Additional spot art by Samuel Hiti and Shutterstock

TABLE OF CONTENTS

Introduction 4
Chemicals 9
Atoms 11
Molecules 13
States of Matter 14
Physical Properties of Matter 18
Chemical Properties of Matter 22
Physical Changes 24
Freezing 26
Condensation 28
Deposition 30
Melting and Boiling 32
Conclusion 36
   Can You Believe It?! 38
   Words to Know 40

There is a glossary on page 40.
Terms defined in the glossary
are in type **that looks like this**
on their first appearance.

INTRODUCTION

Oh, hi! Sorry, I didn't see you.

I was just thinking about how smartphones are such amazing devices. Do you ever wonder why they're made out of particular materials?

If you made a phone out of wood, it could not conduct the electric current needed to work.

Hnngh!

If you made it out of solid lead, it would conduct electricity too well! It would also be extremely heavy—and dangerous!

Phonemakers choose materials carefully to make their phones light, safe, and powerful.

BATTERY

5

But how do they know what materials to choose? They study the **properties** of matter!

Matter is a word for all of the materials that make up the world.

Matter has many forms. A substance is any particular form of matter...

...like the copper inside this phone! (It's a good conductor of electricity.)

Or, the expanded polystyrene beads inside this bean bag chair! This lightweight, cushiony material feels like sitting on a cloud.

Panel 1: Chemistry is the study of substances and how they behave under different conditions.

Panel 2: Let's investigate the basics of chemistry together.

Panel 3: Allow me to introduce myself: I'm Matter!

Panel 4: I'll be your tour guide for this journey.

CHEMICALS

"The word "chemicals" might sound scary, but have no fear!"

CHEMICALS

"A chemical is any of the substances that make up the world's materials."

"The very air you breathe is a combination of the chemicals oxygen and nitrogen and other gases."

SNIFF!

WHUMP!

HOARD-N-SAVE

In fact, everything on Earth is made of chemicals, from table salt to televisions.

ATOMS

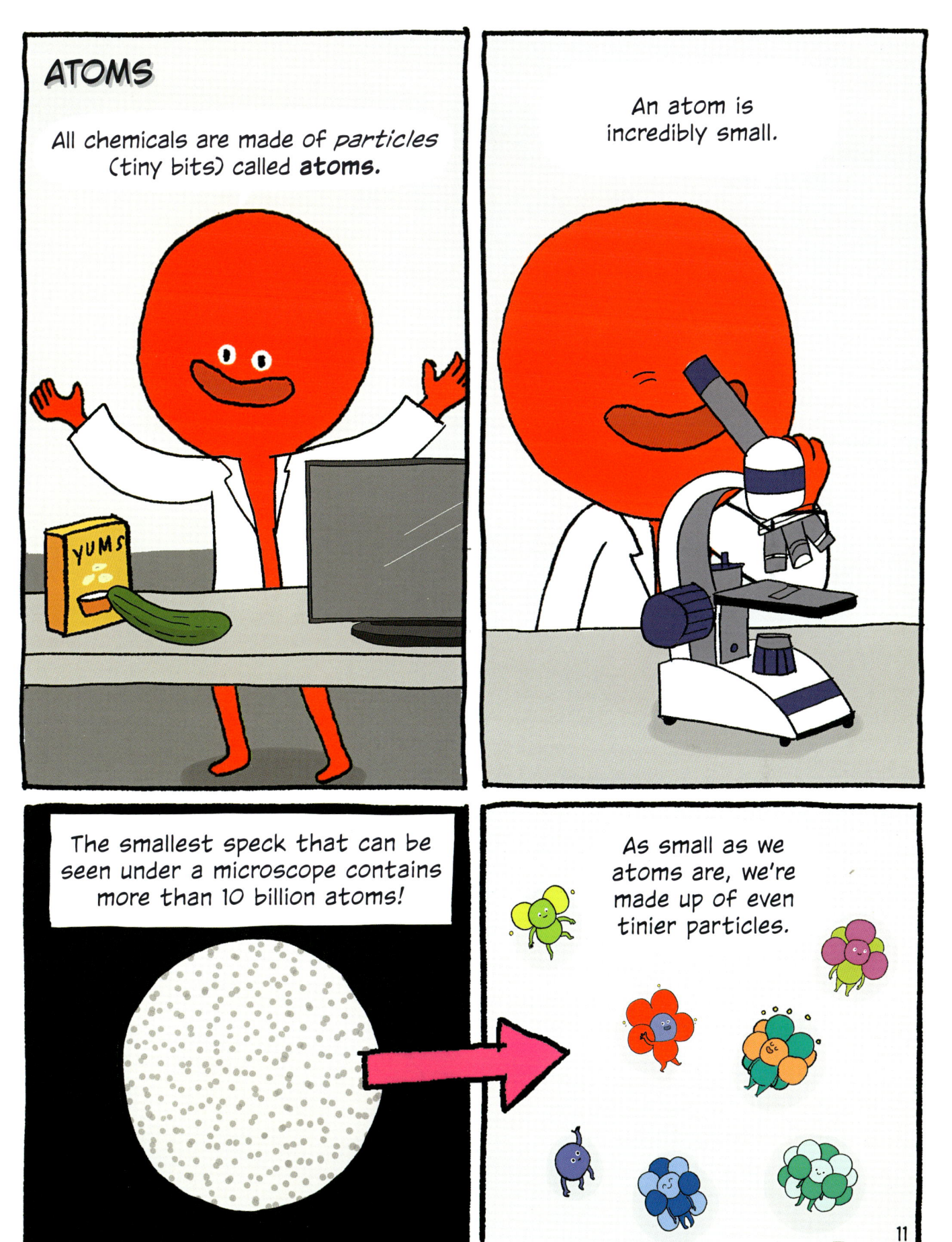

All chemicals are made of *particles* (tiny bits) called **atoms.**

An atom is incredibly small.

The smallest speck that can be seen under a microscope contains more than 10 billion atoms!

As small as we atoms are, we're made up of even tinier particles.

Protons are positively charged particles.

They are found at the center of the atom, called the **nucleus.**

Neutrons are also found in the nucleus. These particles have no charge.

Electrons are negatively charged particles. They move freely around the outside of the nucleus in special zones.

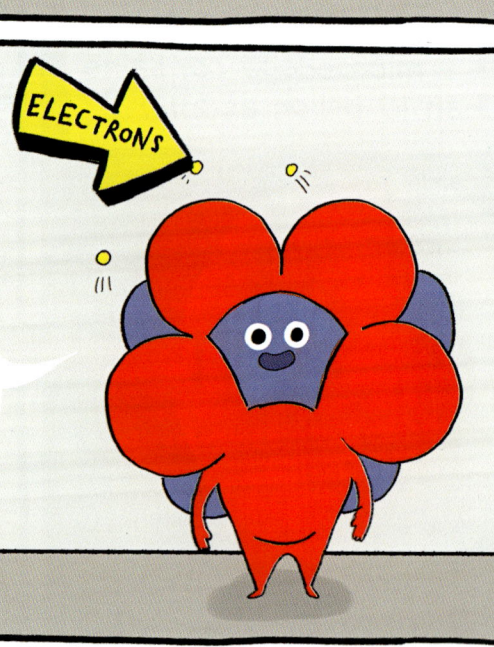

MOLECULES

When two or more atoms join together, they form a **molecule**.

A molecule can be made of a few atoms of the same **element**...

...many atoms of different elements...

...and every combination in between!

13

STATES OF MATTER

THE UNITED STATES OF MATTER

SOLID

GAS

LIQUID

Matter exists in three main states:

solid...

THE UNITED STATES OF MATTER

SOLID   GAS

LIQUID

...liquid...

THE UNITED STATES OF MATTER

SOLID   GAS

LIQUID

...and gas.

Solids have a definite shape.

Their molecules are packed closely together and vibrate in place.

MATTER

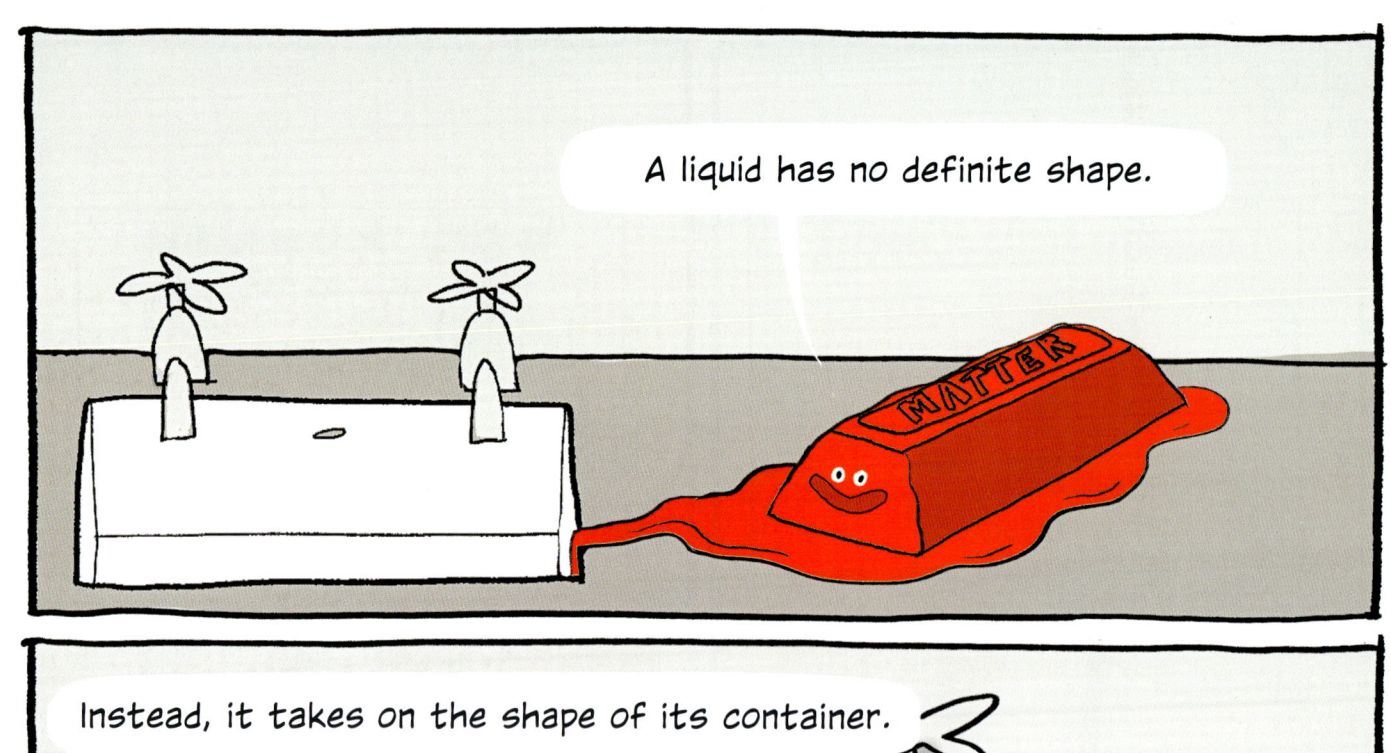

A liquid has no definite shape.

Instead, it takes on the shape of its container.

The molecules of a liquid are not locked in place, so they can move freely past one another.

Like liquids, gases have no definite shape.
Gases expand to fill their containers.

The molecules of a gas are not in contact with one another. They are always zipping around and crashing into one another.

PHYSICAL PROPERTIES OF MATTER
Different kinds of matter vary in their physical properties. Physical properties are things you can see and measure.
1kg
Take a look at these fruits. You can see or measure their color, size, and shape. Those are all physical properties.
You can also find an object's **weight** and **mass.** The two are not the same thing!
Mass is the amount of matter in an object.

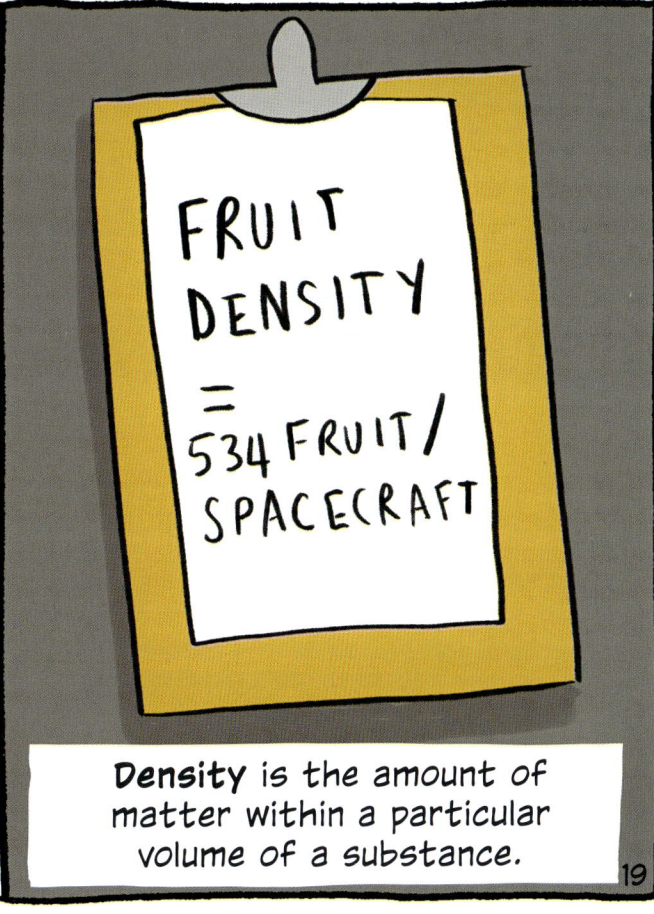

Weight is how much the force of gravity pulls on an object's mass.

An object can be weightless in space, but it still has mass.

Volume is the amount of space matter occupies. This floating fruit is occupying the volume of one spaceship.

FRUIT DENSITY = 534 FRUIT / SPACECRAFT

Density is the amount of matter within a particular volume of a substance.

19

Temperature is a measure of how much internal energy a substance has.

The molecules of a substance may interact differently depending on temperature.

Fruit jams, for example, can be thick at room temperature.

They have a high **viscosity**. Viscosity is a substance's resistance to flow or movement.

The molecules constantly bump into one another and stick together, causing a slow flow.

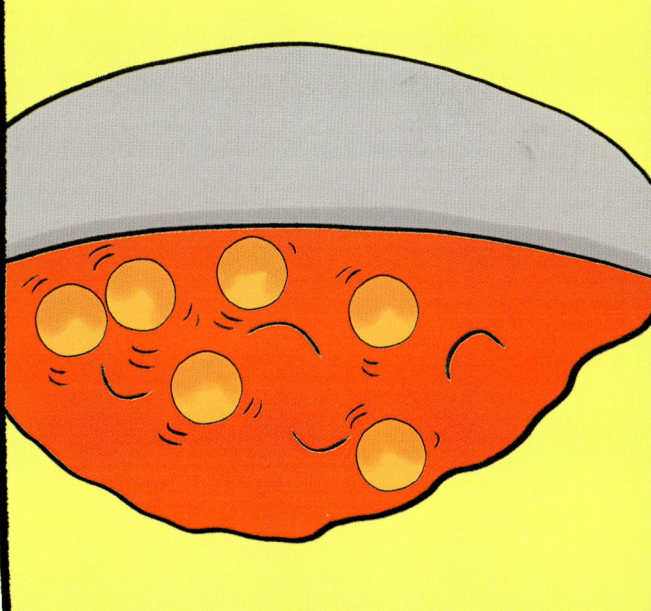

Adding heat energy speeds up the molecules, lowering the substance's viscosity.

The jam flows faster because the molecules can move freely past one another.

21

CHEMICAL PROPERTIES OF MATTER

Physical properties can be identified by simple observation.

But to reveal the **chemical properties** of a substance, a **chemical reaction** must take place.

One chemical property of a substance is its **reactivity**. This is how easily it reacts with other substances!

Some substances react when exposed to oxygen. This property is the ability to oxidize.

For example, oxygen in the air combines with the iron in this horseshoe, changing its properties. This produces a brownish coating called rust.

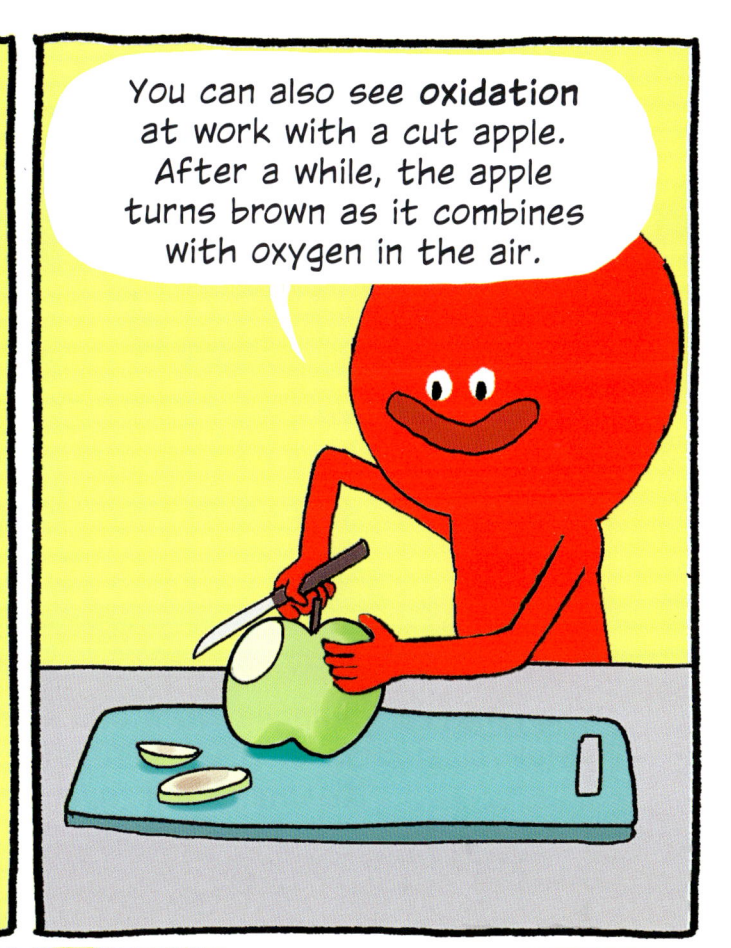

You can also see **oxidation** at work with a cut apple. After a while, the apple turns brown as it combines with oxygen in the air.

Toxicity is another chemical property. A substance that is toxic is harmful to living things.

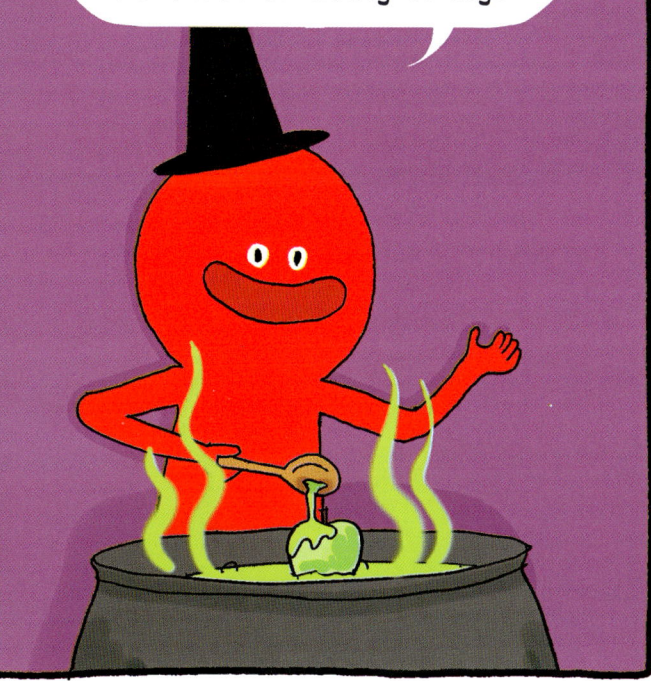

Chemical properties include flammability, or how easily the substance catches fire.

Wood catches fire easily.

23

PHYSICAL CHANGES

The ash is chemically different from the original wood. But matter can also undergo other changes that don't chemically alter it.

When matter changes from one size or state to another, it is called a physical change. When a substance undergoes a physical change, its chemical properties remain the same.

PHYSICAL CHANGE

It's still made of the same substance. Only its appearance has changed.

If you drop glass, it will shatter.

That's a physical change!

Each piece of glass is still made up of the same material.

FREEZING

Some physical changes, such as a change in viscosity or state, can be caused by a change in temperature. **Freezing point** is the temperature at which a substance changes from a liquid to a solid.

When a liquid is cooled, the energy of its molecules decreases. Eventually, the attractive forces between molecules bring them together. The liquid then freezes into a solid.

The freezing point differs from substance to substance. Pure water freezes at a cool 0 °C.

THUNK!

Salty seawater freezes at -2 °C.

Mercury is the only metal that is liquid at room temperature. It has a much lower freezing point than water: -39 °C!

SLOSH!

CONDENSATION

Cooling temperatures can cause a gas to change into a liquid.

That's **condensation!**

CONDENSATION

There's lots of *water vapor* (water in gaseous form) in the air. The water molecules move around freely in warm temperatures.

But in cooler temperatures, the molecules begin to slow down and stick together. As this happens, they change from a gas into tiny water droplets.

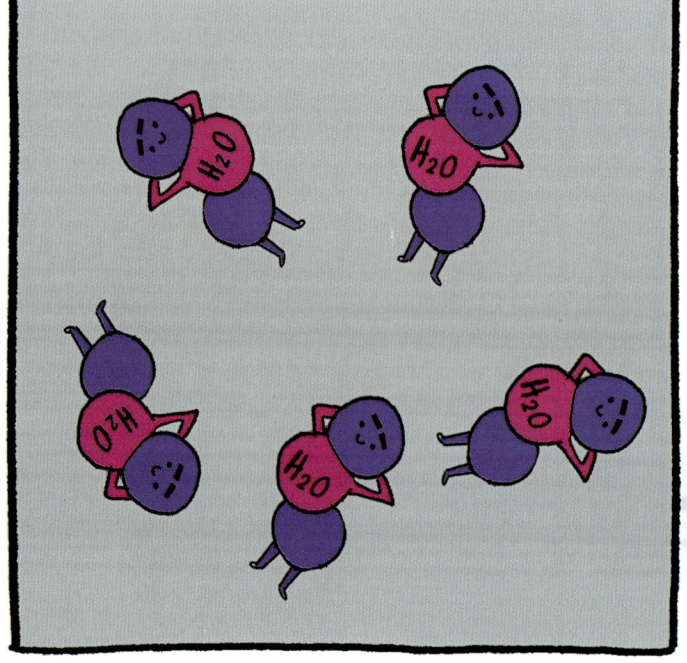

This condensation forms clouds!

If the water droplets get heavy enough, they fall to Earth as rain.

You can see condensation on some mornings. As the warm air hits the cold grass, water vapor in the air condenses on the blades.

The water droplets that collect on the grass are called dew.

29

DEPOSITION

In some cold conditions, gas can skip the liquid phase and turn straight into a solid. This is called **deposition**.

On very cold mornings, you might wake up to frost (frozen water vapor) on the ground instead of dew. Frost often forms through deposition.

During the day, the sun warms the air and the ground.

Once the sun sets, the temperature of the air drops rapidly. Warmer air from the ground rises, while colder air sinks toward the ground.

The water molecules in the cold air quickly latch onto solid surfaces on the ground. They change into a solid-frost, skipping the liquid phase.

Snow is also a result of deposition. When it is below freezing, water vapor within clouds changes from a gas to a solid and falls to the ground.

Occasionally, you can see this process happen in reverse. The sun quickly warms the snow, causing it to change from a solid to water vapor. That's **sublimation** at work!

31

MELTING AND BOILING

Cooling a substance causes its molecules to slow down...

...and heating a substance causes its molecules to speed up!

As a substance is heated, the energy of the molecules overcomes the forces holding them in place.

When the temperature of a solid gets high enough, it changes into a liquid. It's reached its **melting point!**

If the liquid is heated further, it eventually bubbles and changes into a gas. That happens at a temperature called the **boiling point.**

The stronger the bonds between the molecules of a substance, the higher the substance's melting and boiling points.

I'll never let you go.

Water molecules are strongly attracted to one another, so water has a relatively high melting point compared with similar compounds—0 °C. Its boiling point is 100 °C.

The bonds between molecules in gold are much stronger. Gold has a melting point of 1064 °C and a boiling point of 2807 °C!

Panel 1: "As we've seen, changes in temperature can cause a physical change to matter."

CRAK!

Panel 2: "Changes in **pressure** can also cause physical changes—especially to gases. Pressure is the continued action of a weight or force."

Panel 3: "Hngh!"

"Increasing pressure pushes molecules of gas closer together."

This causes the molecules to collide more often, which heats up the gas.

Move!
Ouch!
Watch out!
Hey!

Hngh!

If the pressure increases enough, the molecules stick together and the gas condenses into a liquid.

CONCLUSION

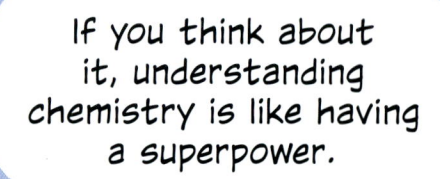

If you think about it, understanding chemistry is like having a superpower.

You can see inside this table...

...or travel through your smartphone...

...or know that the water molecules in this hot tea have more energy than the water in the fish tank.

There's a whole world for you to explore right where you are!

What mysteries will you unravel as you discover the world through chemistry?

CAN YOU BELIEVE IT?!

Nearly all metals appear silver in color. Only **copper** and **gold** do not appear this way.

The word *atom* comes from a Greek word meaning "uncuttable." But **scientists can now split atoms up** into even smaller subatomic particles.

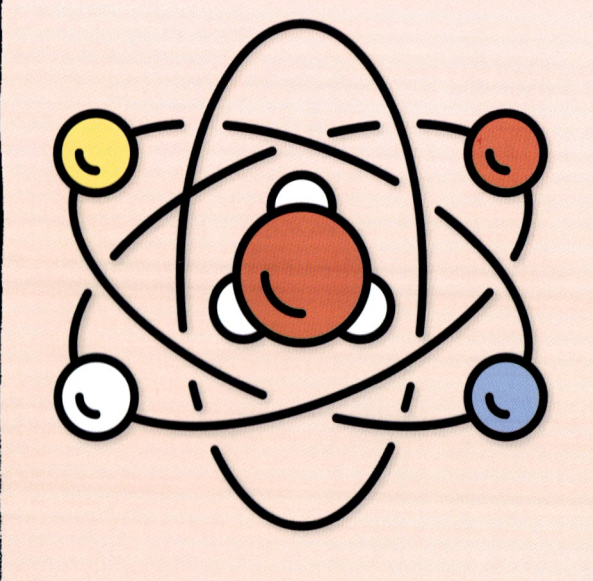

Glass is a kind of matter called an *amorphous solid*—a state of matter that has qualities of both a liquid and a solid!

At incredibly low temperatures, helium gas becomes a superfluid, a fluid that flows with zero viscosity. That means **liquid helium can flow upwards!**

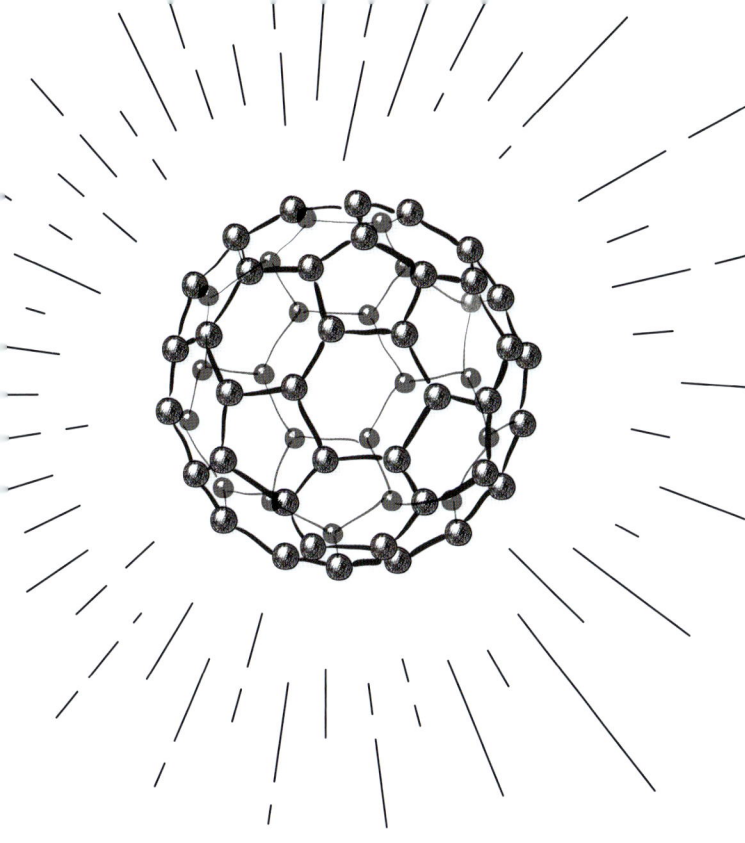

The human-made material *endohedral fullerene*, composed of 60 carbon atoms surrounding a single atom of nitrogen, is the most expensive material ever made. It is so difficult to produce, it costs about

4 billion US dollars per ounce!

There is such a thing as antimatter!

Antimatter resembles ordinary matter but with certain properties of its particles, such as electric charge, reversed. For example, antiprotons are *like protons but with a negative charge.*

In the atmospheres of the planets Uranus and Neptune, where temperatures and pressures are incredibly high,

it rains diamonds!

WORDS TO KNOW

atom one of the basic units of matter.

boiling point the temperature at which a substance turns from a liquid to a gas.

chemical property a characteristic of a particular substance that can be observed in a chemical reaction.

chemical reaction a process by which one or more substances are chemically converted into one or more different substances.

condensation the changing of a gas or a vapor into a liquid by cooling.

density the amount of matter in a particular volume of a substance.

deposition a process in which a gas or vapor changes into a solid without first becoming a liquid.

element a substance made of only one kind of atom. There are 118 chemical elements.

freezing point the temperature at which a liquid becomes a solid.

mass the amount of matter in something.

melting point the temperature at which a solid becomes a liquid.

molecule two or more atoms chemically bonded together.

nucleus the center of an atom. The nucleus is made up of protons and neutrons.

oxidation any chemical process in which a substance combines with oxygen.

pressure continuous physical force exerted on or against a substance.

property a quality belonging specially to something.

reactivity a measure of how well a substance readily combines with other substances.

sublimation process in which a solid changes into a gas or vapor without first becoming a liquid.

temperature a measure of the amount of heat (internal energy) in a substance.

viscosity a measure of the resistance of a fluid (liquid or gas) to flow. Fluids with high viscosity, such as molasses, flow more slowly than those with low viscosity, such as water.

volume the amount of space an object occupies.

weight a measure of the heaviness of an object. The weight of an object is the gravitational force on the object.

+
540 M

Meyer, Cassie,
Chemistry and matter /
Heights NONFICTION
11/22